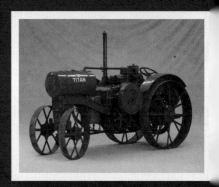

D1633807

BEAUTIFUL TRACTORS

PORTRAITS

of

ICONIC
MODELS

BEAUTIFUL TRACTORS

PORTRAITS

of

ICONIC

MODELS

Courtesy of the Paul Rackham Collection

by RICK MANNEN

photographed by CLIVE STREETER

foreword by STUART GIBBARD

F

FRANCES LINCOLN LIMITED
PUBLISHERS

Frances Lincoln Limited
4 Torriano Mews
Torriano Avenue
London NW5 2RZ
www.franceslincoln.com

British Library Cataloguing in Publication Data
A catalogue record for this book is available from the British Library

This book was conceived, designed and produced by

Ivy Press

210 High Street, Lewes, East Sussex, BN7 2NS, UK

Creative Director **Peter Bridgewater**
Publisher **Jason Hook**
Art Director **Wayne Blades**
Senior Editor **Jayne Ansell**
Designer **Ginny Zeal**
Photographer **Clive Streeter**
Illustrator **David Anstey**

ISBN: 978 0 711233 546
First Edition: 2013
Printed in China

Colour origination by Ivy Press Reprographics
9 8 7 6 5 4 3 2 1

CONTENTS

Foreword 7

Introduction 9

THE TRACTORS *16*

REPORTAGE *98*

Glossary 110

Shows & Associations 110

Index 112

FOREWORD

THE THREE FACTORS THAT HAVE HAD PROBABLY the greatest influence on modern agriculture during the past century are plant breeding, chemical pesticides and the farm tractor. The first two, although undoubtedly important – if somewhat controversial – breakthroughs, are difficult to romanticise, but the tractor is different: it is an object of passion for an ever-growing band of dedicated enthusiasts.

Tractors, working the fields and bringing home the food that we eat, have become part of the fabric of the countryside. Almost all of us who live in a rural community have been touched by their presence at some time in our lives – whether by learning to drive on an archetypal 'grey Fergie', or seeing the latest John Deere flash past our windows. Their influence also extends to the towns and cities, where industrial tractors have carried out a multitude of tasks, from shifting refuse to clearing snow.

I grew up on a farm, and as a child my interest in machinery was aroused as I 'farmed' the carpet with my fleet of toy tractors. When I progressed to driving the real thing I revelled in learning the intricacies of maintenance and the skills of ploughing. Many years later I changed direction and moved into journalism – researching and writing about tractors, of course!

My interest has never diminished, and I have been collecting vintage tractors for almost 40 years now. In that time I have seen the hobby grow and grow; not just in the UK, where I live, but also across North America, Ireland, mainland Europe and Australasia. It is an inclusive pastime, encompassing enthusiasts from almost every walk of life – from farm workers and agricultural students to doctors, dentists, stockbrokers and even actors from the world of film and television.

The popularity of the hobby has seen collections spring up throughout the world. In Britain, one of the most comprehensive is that of Norfolk collector Paul Rackham. Paul has dedicated himself to gathering together a wonderfully diverse selection of tractors, and many of the gems in his collection have been captured by the stunning photography in this book. The machines are brought alive by the informative text penned by my great friend and fellow historian, Rick Mannen. Read on and discover what it is that fascinates us about these fabulous beasts.

Collecting tractors is almost certainly motivated by a degree of nostalgia, but there is more to it than that. A tractor is raw, mechanical power in its purest form – much more than a mere collection of gears and cogs, castings and steel. It has a heart – a life – and is truly a beautiful machine.

As Harry Ferguson once said, 'Beauty in engineering is that which is simple, has no superfluous parts and serves exactly the purpose.'

Stuart Gibbard, November 2011

INTRODUCTION

BEAUTIFUL TRACTORS CONTINUES A POPULAR series that, up until now, has concentrated on champion animal breeds. However, a vintage tractor – as anyone who has operated one will know – can often display the personality, good or bad, of a farm animal. Many a farmer has praised or cursed his tractor, but at the end of the working day put it to stable, tenderly wiped it down and otherwise coddled it in the same way as a favourite draught horse. The tractor was, after all, one of his largest investments. A farmer could feel proud whilst working his fields with a fine-looking tractor. It represented class, power and modernity, and quickly became more than just another farm tool. Tractors gained a central role in agriculture and have entered into the imaginations of the people who still enjoy them today.

In these pages you will find an exceptional selection of 20th-century tractors, which are part of the collection of Paul Rackham. Beauty is in the eye of the beholder and the tractors presented here offer a range of styles, from primitive, hulking antiques to the classic, streamlined gems of the post-war era. Accompanying each superb photograph are details on the tractor's history, features, uses, related models, size and distribution.

This variety in design deepens the appeal of the vintage tractor to people today. How many grown men will remember having seen a thundering Oil Pull, a streamlined Allis-Chalmers or a popping John Deere in their younger days and dreamt of being able to operate one of these marvellous beasts? Tractors became objects of interest for collectors after the Second World War as people grew nostalgic for the quieter times of the past. At that time old tractors could be had for little more than scrap value, or just for the effort of taking them away. By the 1960s more and more people began to realise that the supply was dwindling, and the machines started to gain some monetary value. In some areas most of the early tractors were taken by the wartime scrap drives, but with the coming of the age of containerised shipping, collector tractors can now be shipped worldwide and enthusiasts delight in adding a sought-after, unusual model to their collection.

Above: From vintage to modern machine, tractors represent an impressive display of engineering prowess.

DEMAND FOR TRACTORS

THE TRACTOR ERA IS LITTLE MORE THAN A SHORT chapter in the overall history of agriculture. In the earliest days, humans hunted the abundant wild game and gathered from nature's bounty of fruits, nuts and vegetation. But as populations rose and distinct societies were formed, people saw a need to secure more dependable sources of food and began growing crops. Popular belief places the first cultivation of crops in the 'Fertile Crescent' of today's Middle East, but there is plenty of evidence that the growing of crops started spontaneously in many parts of the world. For many centuries human and animal labour were the only means of tilling the soil. Reaping and bringing the grain to the bins took days or weeks of hard, back-wrenching work using hand tools.

The Industrial Revolution of the 18th and 19th centuries was a remarkable coming together of human progress and invention. Innovations in metallurgy, steam power and advanced mechanisation began to transform farming methods that had remained constant since biblical times. But all of these new machines came with a price, and that

Above: The introduction of steam power transformed farming methods that had served since biblical times.

price was a hunger for power. That James Watt's steam engine of the late 1770s would benefit the farmer was not immediately evident, as it was designed to power the machinery of factories and mills and was simply too large for use on the farm. The use of the smaller, portable steam engine blossomed after the late 1700s and the traction engine arrived by the middle of the 19th century. In Britain, one of the first recorded uses of steam in threshing was in Yorkshire in 1799.

In the Americas, steam threshing did not come into vogue until the 1850s. The vastness of the mid-North American plains called for appropriate equipment. By the turn of the 20th century, American, Canadian and British builders were producing large, powerful traction engines that were specifically designed for heavy ploughing service. But the age of steam on the farm would prove to be short-lived. In some of the plains areas, such as the treeless expanses of Saskatchewan in Canada, fuel for the big steamers had to be brought in and sour water played havoc with the boilers. Another sort of farm power was needed.

THE PIONEER TRACTORS

JUST AS EARLY CULTIVATION HAD SPRUNG UP IN disparate parts of the world, so too did the internal combustion engine – or 'explosive' engine as it was called in the early years. The big leap to a practical design is credited to Nikolaus Otto in Germany in the late 1870s, with his four-stroke compression and spark-ignition motor. Another name of great importance is that of Rudolph Diesel, another German whose 1890s patents led to the engine and the type of fuel named after him. These men created designs that are still in use today.

The Charter Gas Engine Company of Illinois was one of America's pioneer builders, and seized on the notion of applying internal combustion power to traction engines as early as 1889. A number of companies jumped on the bandwagon, and names such as Case and International Harvester are as familiar today as they were in those early years. John Froelich built a tractor in 1892 that is a direct antecedent of the famed John Deere line.

The pioneer tractors were built with the plains farmer in mind. Of stout design with heavy channel frames, they were mainly suitable for belt work such as threshing and were not yet ready for heavy ploughing jobs. Most of them had single- or two-cylinder motors that were designed to start and warm up on petrol and then work on kerosene or other less expensive fuels. In those days kerosene was half the price of petrol, a considerable saving over a day's work. Although they worked well enough, they were not enduring designs and suffered from ignition, carburetion and transmission components that had not yet evolved sufficiently to make these pioneer internal combustion traction engines fully reliable.

In the first years, such machines were referred to as petrol (gasoline in the USA) or oil traction engines. The American company Hart Parr, of Charles City, Iowa, is credited with the first use of the word 'tractor' in describing their creations of 1902. The word tractor probably derives from the Latin *trahere*, meaning 'to pull', but it may have been shortened from 'traction engine'. The Hart Parr company was formed specifically to produce internal combustion tractors and wanted to set itself apart from the old steam firms.

Above: John Froelich's pioneering tractor of 1892 is a direct antecedent of the John Deere line.

EARLY TRACTOR DEVELOPMENT

IN THESE EARLIER TIMES A GREAT MANY OF THE products reaching the farmer were of such poor quality that they were of little use. Farmers simply did not have the money to throw away supporting some dreamer's experiments. It was clear that certain industry standards had to be adopted. At first prospective buyers had to rely on the honesty of the manufacturers and their advertised horsepower ratings. Then in 1908 the Winnipeg Motor Competitions developed some of the first scientific testing protocols, employing qualified observers and instruments. This helped to separate the wheat from the chaff, kept the builders honest and provided buyers with accurate information. This type of testing was eventually formalised at the University of Nebraska, USA, and any tractor maker wishing to have success in North America was obliged to pass their tractors through these tests.

Many companies concentrated on building tractors for the heavy work of breaking the tough virgin sod on the millions of acres of new lands being opened for settlement around the world. These were the huge 'prairie tractors'

Above: Early tractor design reflected the need for strong heavy machines that could work new lands for settlement.

intended to do the work formerly undertaken by steam engines. They were no longer the unreliable experiments of the pioneer builders but were strong, heavy, ready to work and able to last! Costing over $3000 (about £2000), they were beyond the means of the average farmer and intended for the custom ploughman and thresherman.

Hart Parr, Rumely, International Harvester and Kinnard-Haines were some of the better-known American companies. Sawyer-Massey and Goold Shapley & Muir were successful Canadian builders. Well-known steam traction builder Marshall, Sons & Co. of Gainsborough, England, built heavy tractors for prairie work in Canada and elsewhere. There were builders on the Continent as well. These prairie giants found their way to the open steppes of Russia and Ukraine, to southern Africa, Australia, Argentina and elsewhere.

There were also the first signs of some diversity in tractor design. Saunderson of Bedford, England, began offering lighter models for use on the smaller, more developed farms of eastern North America, England and mainland Europe.

TRACTORS & THE GREAT WAR

SOON, THE PRAIRIE LANDS HAD BEEN BROKEN TO the plough and the big tractors were relegated to threshing duties. But in 1914 much of the world found itself at war. Far too many young men were taken from the farms and into military service, creating a serious shortage in the workforce needed to produce the badly needed crops. This created a demand for a new type of tractor, something light and inexpensive that could undertake a variety of jobs formerly done by horses. The tremendous demand for tractors resulted in hundreds of companies around the world suddenly being engaged in building them with the encouragement of their governments.

American automotive pioneer Henry Ford worked with the British, Canadian and American governments to create a new type of tractor to assist with wartime food production. The Fordson had a simple design and was built using Mr Ford's assembly-line techniques, and was therefore relatively inexpensive. It featured what became known as the 'unit frame' design, an innovation that greatly reduced the weight and costs

of tractors. Earlier tractors had used heavy structural steel frames to carry the motor and transmission components, whilst the Fordson used the motor and transmission castings themselves as the main frame. Other companies were forced to offer similar designs.

With a price range of $1000 (about £650) and under, tractors could now be made available to average farmers around the world. Farmboys went to war in the age of horses and came home in the age of aircraft. Many had operated motorised equipment for the first time and had learnt how to maintain and repair it. Few of them had the appetite to return to the farm and spend their days working with old horses. To keep the lads on the farm father was obliged to consider modernising!

The First World War sped up technological development dramatically. It saw the development of military tanks and other tracked vehicles and this immediately caught the attention of tractor builders. Many of the automotive manufacturers in Europe entered the tractor market at the end of the war and the industry took on a greater international flavour.

Above: The unit frame design of the Fordson, introduced by Henry Ford, heralded an era of lightweight tractors on the farm.

MODERN TRACTORS

MAJOR INNOVATIONS CAME TO THE TRACTOR industry in the years between the two world wars. Amongst these were the common use of diesel motors, pneumatic tyres, power take-off systems, the Ferguson hitch and other hydraulic systems. Some of the first successful four-wheel drive tractors also arrived on the scene. The Second World War brought greater advances in technology. Many of the companies that built tractors before the war found themselves building highly engineered products such as aircraft for the war effort. When the war ended these companies could draw on the experience of their newly skilled engineers to develop the more sophisticated designs that were now being demanded by farmers.

Following on the heels of the streamlining craze in transportation, many companies – including International Harvester, John Deere, Allis -Chalmers and Cockshutt – contracted with well-known firms such as Raymond Loewy and associates of the Society of Industrial Design to develop exciting new tractor stylings. This group had its roots in the Art Deco and Bauhaus movements of the 1920s, whose

Above: Electronic systems in the cab are increasingly used by today's designers to control sub-systems such as the hydraulics.

philosophies encouraged combining style with utilitarianism in everything from household toasters to railway locomotives – just because it is useful doesn't mean it can't look good too!

Creature comforts such as enclosed cabs soon became the norm and one could have a radio, heater and even air conditioning. By the early 1960s tractors were beginning to push the boundaries past 100 horsepower and the race was on to make them bigger and more powerful. Even farm equipment succumbed to the age of computerisation and advanced electronics. Under the hoods and inside the cabs, today's tractors offer fingertip control of speed shifting and hydraulics. The operator can be linked up to satellite positioning technology that allows him or her to sit back in a comfortable upholstered seat in a climate-controlled, dust- and soundproof cab as the stereo system hums away and the tractor guides itself faithfully along the rows. The tractor industry of the 21st century is, more than ever, an international affair, with worldwide production now including a wide range of tractors arriving from Asia.

COLLECTING TRACTORS

THE SCRAP DRIVES OF THE SECOND WORLD WAR cleaned up lots of the antique tractors and people began to worry that they all might disappear. As early as the 1940s a few forward-looking people began saving the old equipment for public museums and private collections. Some diehard threshermen oiled up their old charges and started holding reunions. From a few such events grew an entire movement of tractor preservation, restoration, operation, associations and shows. Tractor shows have spread throughout the world and range from small local events to huge gatherings spread over several days. Working events are becoming more widespread as new generations want to see the older tractors put through their paces threshing or working the fields. Tractor pulls have also become popular, with comparable tractors hauling measured loads for fun or in competition.

Restoration is conditional on the resources of the owner, his or her mechanical skills, budget and preferences. Some collectors use their tractors chiefly for parades and display and wish to have full cosmetic restoration. If the overall condition warrants, there is a newer school who prefer to leaving the tractors as they found them, with their original, working appearance. Others may wish to do heavier mechanical repairs in order to be able to work them hard. Collectors' interests change with the passage of time and, with the early prairie tractors now commanding six-figure prices, many collectors are seeking out the 'latter-day classics' they might have known when they were growing up. There is also a growing following for the smaller, garden-type tractors, as some of them now date back a half century or more.

The larger tractor shows offer judging and prizes, but generally this has never been as formal as it is for automotive or champion farm animal judging. Where it is formalised, judges will look for the use of authentic replacement parts and fittings and the overall quality of the mechanical and cosmetic restoration. More often, the 'best in show' award might be made to someone who has been a long-time show member or supporter. But for most collectors the pleasure is in joining with others to enjoy their beautiful tractors.

Above: Tractor shows and working events give collectors and restorers the opportunity to admire and compare their machines.

THE TRACTORS

Can one *fall in love* with a FERGUSON at first sight or become as attached to an *Allis-Chalmers* as to an adored animal? How *dear* can a John *Deere* be? The answers are a RESOUNDING yes, yes and *immensely*. Collectors and restorers of vintage tractors love their steel workhorses *deeply, madly and truly*. These pages show you why...

FERGUSON-BROWN, MODEL A

UK, 1937

Harry Ferguson of Belfast, Northern Ireland, developed his groundbreaking Ferguson System hydraulic hitch and draft control and in 1933 tested it with the Ferguson-Black. In 1936 he collaborated with the David Brown Company to put the system into production as the FERGUSON-BROWN, or MODEL A tractor. Production lasted until 1939 when Ferguson went to America to join forces with Henry Ford.

Features

The Ferguson-Brown initially sported a Coventry-Climax four-cylinder petrol motor but switched to a David Brown motor from *c.*1937. It had a three-speed transmission and was the first production tractor to feature the Ferguson System advanced hydraulics and three-point hitch with draft control. Ferguson evidently had a soft spot for the colour grey.

Uses

The Ferguson-Brown was designed as a light ploughing tractor. It had a variety of three-point hitch-mounted implements and could be fitted with a pulley attachment to run belt-driven machinery. It is a valued collector's item today.

Related Models

The Ferguson-Brown evolved into the popular Ford-Ferguson and, later, the TE-20 Ferguson. David Brown used its experience with Ferguson on its own VAK 1 tractor of 1939.

Power & Size

20 hp; weight: 839kg (1850lb)
length: 287cm (113in)
width: 147cm (58in)
height: 119cm (47in)

Manufacturing & Distribution

This model was produced by David Brown's Park Gear works in Huddersfield with Harry Ferguson handling the marketing. Costing almost twice as much as the popular Fordson, only 1350 units were built. They were sold in Britain.

Huddersfield, England

ALLIS-CHALMERS, MODEL B

USA, 1939

Allis-Chalmers (A-C) was based in Milwaukee, Wisconsin, and dated back to the turn of the 20th century. Its tractor production began in 1914 with some sensible, lightweight models, and the company quickly grew to become one of the world's largest producers of farm machinery. The MODEL B was one of A-C's most popular and durable models, designed for the small farmer and vegetable grower.

Features

The Model B used a four-cylinder motor fuelled by petrol or distillate. A diesel version was built in Britain. A novel weight and cost-saving 'torque tube' frame mated the motor to the three-speed transmission. The A-C tractor line was dressed in a solid, bright 'Persian Orange', with modernistic sheet metal conceived by American industrial designer Brooks Stevens.

Uses

Excellent ground clearance and adjustable wheel widths made the Model B perfect for the market gardener, with a range of matching mountable tillage equipment available. The B can still be found hard at work today and is a popular sight at shows.

Related Models

A row crop version, the Model C, of essentially the same capacity appeared in 1940. A slightly larger EB model was built in Britain.

Power & Size

16 hp; weight: 1021kg (2250lb)
length: 279cm (110in)
width: 132–157cm (52–62in, adjustable)
height: 196cm (77in)

Manufacturing & Distribution

Over 120,000 models were produced in Milwaukee, Wisconsin from 1937 to 1957. Examples can also be found in the UK, mainland Europe, Australia and New Zealand.

Wisconsin, USA

JOHN DEERE, MODEL R

USA, 1954

One of the first John Deere-designed tractors, the Model D was introduced in 1923. In 1949 that venerable tractor was updated with the MODEL R, the company's first diesel tractor and their largest up to that time. When put through the Nebraska tests, the R proved to be one of the most fuel-efficient tractors available. Professional styling from the industrial designer Henry Dreyfuss gave it a touch of class.

Features

The two-cylinder, horizontal diesel engine was started using a smaller, two-cylinder, petrol 'pup' motor. It had a five-speed transmission, live power take-off and John Deere's 'Powr Trol' hydraulic lift system. It came in John Deere Green with the company's own blend of yellow on the wheels.

Uses

The Model R was the first 'big' tractor for many farmers and excellent for all kinds of drawbar work or to run a good-sized thresher. With its distinctive two-cylinder beat, the big R is now a prized attraction at shows and is hard to beat at the vintage tractor pulls.

Related Models

The R was succeeded by the Model 80. The similar 720, 820, 730 and 830 models closed out the two-cylinder diesel era by 1960.

Power & Size

43 hp; weight: 3447kg (7600lb)
length: 373cm (147in)
width: 202cm (79.5in)
height: 198cm (78in)

Manufacturing & Distribution

The Model R was manufactured at the company's works in Waterloo, Iowa from 1949 to 1954, with just over 21,000 units built. It was especially popular in North America and Australia but also sold in Britain, mainland Europe and New Zealand.

Iowa, USA

FERGUSON, MODEL TE-20

UK, 1947

When his partnership with Henry Ford turned sour after the Second World War, Harry Ferguson returned to Britain and built the MODEL TE-20, first released in 1946. But he also established a factory in Detroit, Michigan; the US version, the TO-20, appeared in North America in 1948. In 1953 he sold the company to Massey-Harris Co. of Toronto, Canada, which was eventually to become Massey-Ferguson.

Features

The first TE-20 had a four-cylinder Continental petrol motor; a later version used a Standard Motor Co. power plant. An optional diesel motor was fitted for British dealers. Attachments included a hay mower and the outfit was completed with a 'game flusher' to warn animals of impending danger. The 'Fergie' was painted in the colour that would become known as Ferguson Grey.

Uses

The TE-20 was rated as a two-plough tractor and, with the innovative Ferguson System three-point hitch with depth control, and four-speed transmission, it worked reliably in most conditions. An optional rear-mounted belt pulley increased its versatility.

Related Models

The Ford-Ferguson 9N and 8N were competing tractors of similar capacity. The Ferguson Twenty80 and Twenty85 were similar models aimed at the Canadian market.

Power & Size

24 hp; weight: 1089kg (2400lb)
length: 292cm (115in)
width: 163cm (64in)
height: 142cm (56in)

Manufacturing & Distribution

The TE-20 was built at the Standard Motor Company's works in Coventry from 1946 and enjoyed excellent sales in the UK, North America, Europe, Australia, New Zealand and southern Africa.

Coventry, England

ALLIS-CHALMERS, MODEL U

USA, 1933

The MODEL U was a pioneer tractor for its use of pneumatic tyres. In 1933, in concert with the Firestone tyre company, A-C hired American racing driver Barney Oldfield to tour autumn fairs in America with a Model U, where he reached the unheard-of speed of 64 mph. A-C was renamed Deutz-Allis in 1985 and became part of the AGCO farm equipment empire in 1990.

Features

The first Model U used a four-cylinder Continental motor; a later version had an A-C motor. It featured four forward speeds. Its appearance was utilitarian with a unit frame and minimal styling but it was dressed out in solid Persian Orange. It could be purchased with steel wheels and some of the first in production had the name United cast into the top radiator tank.

Uses

The Model U was adaptable to many types of drawbar work and on rubber tyres was quite useful in haulage. A standard side-mounted belt pulley was perfect for threshing and grain grinding. Rarely in daily use now, the history-making Model U can be a star attraction at a tractor show or ploughing match.

Related Models

Although it compared favourably to International Harvester's 15-30 model, the U was really the last in the line of unstyled tractors for A-C.

Power & Size

30 hp; weight: 2087kg (4600lb)
length: 301cm (118.5in)
width: 160cm (63in)
height: 136cm (53.5in)

Manufacturing & Distribution

First built in 1929–30 for the United Tractor & Equipment Co. of Chicago, Illinois, some 10,000 were made for A-C's own sales department until c.1944. Sales covered North America, the UK, Europe, Australia and New Zealand.

Illinois, USA

FRENCH AUSTIN, MODEL R

FRANCE, 1928

Herbert Austin began designing tractors at his automobile plant in Birmingham in 1918, and the first production models appeared the following year. Also in 1919 he opened a plant in Liancourt, France, which handled the bulk of the tractor production, including the MODEL R. The French and British models were initially almost identical, with parts built in England for assembly in Liancourt.

Features

The Model R motor was adapted from the Austin car engine to use petrol or kerosene. The French-built tractor had three forward speeds. Unit frame construction was employed with some refinements such as sheet metal covering the delicate motor accessories. Paint schemes included dark green with red wheels (shown here) and light blue with red wheels.

Uses

The Austin was meant to compete with the Fordson as a light, all-purpose tractor on all forms of drawbar and belt work. As a display attraction, a well-tuned Austin still delivers a nice turn of ploughing today.

Related Models

The SA3 and DE30 were successor models from the Liancourt works. The R was also built in vineyard and industrial versions.

Power & Size

20 hp; weight: 1406kg (3100lb)
length: 279cm (110in)
width: 155cm (61in)
height: 140cm (55in)

Manufacturing & Distribution

Production continued in Birmingham until about 1932.
Production in Liancourt, France outlasted that in Britain.
The R was sold in the UK, Europe, Australia and New Zealand.

Birmingham, England   Liancourt, France

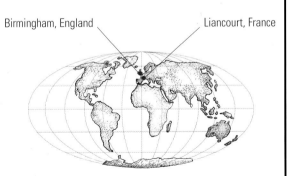

CATERPILLAR, MODEL D2 (5J)

USA, 1942

Benjamin Holt conceived the Caterpillar name for his tracked tractors and merged his company with C.L. Best to form Caterpillar in 1925. Crawler tractors came of age in the First World War after the success of the battle tank, and the successful MODEL D2 honoured that heritage. The Caterpillar name has endured throughout the age of mergers and corporate takeovers.

Features

The D2 was the company's smallest diesel tractor, with a four-cylinder main motor that had to be started with a two-cylinder petrol auxiliary motor, sometimes called a 'pup' or 'donkey'. It featured a five-speed transmission. The 5J designation referred to its 50-in (127-cm) tread gauge. The distinctive Caterpillar Yellow colour has now entered the public consciousness, but 'Cat' crawlers were never meant to be fancy — merely powerful.

Uses

The robust D2 was designed for heavy work and found many roles in agriculture, construction and the military. There is a new-found love for old Cats today, but those good diesel motors are still capable of honest hard work.

Related Models

The D series came in five sizes, up to the powerful D8. The R2 was similar but with a petrol/kerosene engine. The D2 (3J) had a 15.75-cm (40-in) tread width.

Power & Size

30 hp; weight: 3107kg (6850lb)
length: 272cm (107in)
width: 168cm (66in)
height: 147cm (58in)

Manufacturing & Distribution

Built in Peoria, Illinois, the D2 enjoyed strong sales in North and South America, the UK, Europe, Australia and New Zealand. It was produced from 1938 until 1957.

Illinois, USA

JOHN DEERE, MODEL 3130

GERMANY & SPAIN, 1975

John Deere had been an established name in tractor building in the USA since 1918. In 1956 acquisition of the Heinrich Lanz tractor works in Germany gave the company a window to the European marketplace. The MODEL 3130 complemented the 30 series tractors being built at home in Waterloo, Iowa.

Features

The 3130 featured an excellent six-cylinder diesel engine and had 12 forward speeds. Four-wheel drive was an option, along with a factory-built cab. Decked out in traditional John Deere Green with John Deere Yellow wheels, the 'beetle-brow' hood design of the pictured tractor denotes an earlier version, as the front cowling was swept back on later units. The advertising declared, 'Nothing Runs Like A Deere!'

Uses

Built as a higher horsepower, mid-range tractor, this gutsy machine with a good speed selection and ultra-modern hydraulics was capable of a full range of chores. Not built in high numbers, the 3130 is coming into its own for collectors and admirers of the latter-day classics.

Related Models

The 3130 was part of a wide selection of sizes in the 1970s sometimes referred to as Generation II or the 30 series. The 2840 model is a similar version built in Germany from 1977 to 1979.

Power & Size

97 hp; weight: 4218kg (9300lb)
length: 406cm (160in)
width: 203cm (80in)
height: 249cm (98in)

Manufacturing & Distribution

Production was shared between the factories at Mannheim, Germany and the Iberica Works in Getafe, near Madrid, Spain from 1973 until 1979. This design was targeted at the Canadian market although some were sold in Britain and mainland Europe.

Madrid, Spain

Mannheim, Germany

FORDSON MAJOR, MODEL E27N

UK, 1950

Fordson moved its tractor production to England in 1933. Its Model N, though very successful, was becoming rather dated; the MODEL E27N had many improvements but development had been impeded by the war. Fordson gave the E27N the name Major to set it apart from its ancestors and highlight its design changes. Even as an interim model, it proved to be a worldwide success.

Features

The Major was dark blue, a colour that would endure through much of Ford's later tractor production. A three-speed transmission and modern final drives were mated with a hydraulic lift hitch system to make a fine working unit. A Perkins diesel motor was offered as a popular option. Some Majors sported front lights on wide brackets and farmers affectionately called them 'Longhorns'.

Uses

Versatility and reliability was the Major's hallmark. With a higher ground clearance than the old N series and the ability to carry mounted implements, it found many uses in row crop farming and easily handled traditional chores such as threshing.

Related Models

The E27N gave way to the E1A (New Major) and the Ford Dexta, and was an integral link in the evolution of the modern Ford tractor line.

Power & Size

30 hp; weight: 1860kg (4100lb)
length: 338cm (133in)
width: 165cm (65in)
height: 208cm (82in)

Manufacturing & Distribution

The Major was built in Dagenham, Essex from 1945 to 1952 and enjoyed sales across the UK, North America, Australia, New Zealand and southern Africa. Nearly a quarter of a million were built, a testament to the farmer's desire for modern tractors after the cessation of the Second World War.

Essex, England

FORDSON, MODEL F

USA, 1917

The innovative Fordson MODEL F was built using Ford's cost-saving assembly-line techniques. The design for the Model F was developed with the assistance of the Board of Agriculture in Britain as an effort to create an inexpensive light tractor to boost wartime food production. The first Model Fs were identified as Ministry of Munitions or MOM tractors.

Features

The four-cylinder motor, three-speed transmission and worm gear final drive components were assembled to make up the unit frame, saving both weight and expense. The early MOM series tractor was distinguished by its so-called 'ladder side' radiator, while later, a solid-sided radiator was made standard. The series was tastefully decked out in basic grey with red wheels.

Uses

The Fordson was a light ploughing tractor with two-drill capacity. It was also excellent for most other light field and belt work. Other companies began making implements that could be used with the ubiquitous Fordson. The early MOM version is an especially sought-after collector's item today.

Related Models

The slightly more powerful Model N was produced in Cork, Ireland, in 1928 and later in Dagenham, England, leading to the well-known Fordson Major tractors.

Power & Size

10–20 hp; weight: 1225kg (2700lb)
length: 259cm (102in)
width: 160cm (63in)
height 140cm (55in)

Manufacturing & Distribution

Well over half a million units were built from 1917 to 1928 at Dearborn, Michigan. The British government distributed the initial 6000 units. They were also sold across the USA, Canada, Australia, New Zealand, Europe and southern Africa.

Michigan,
USA

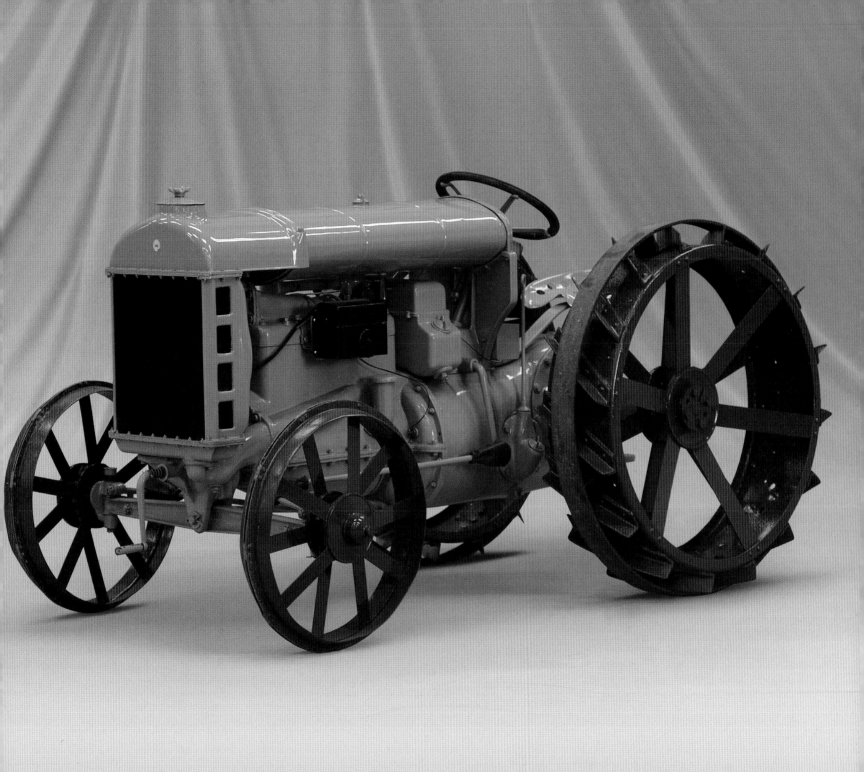

FORDSON, MODEL N
UK, 1938

Henry Ford had a place in his heart for his Irish ancestry and moved tractor production from Michigan to Cork, Ireland, in 1928. The MODEL N emerged from the Cork works, replacing the venerable Model F. However, in 1932 Model N production was relocated to Dagenham, England, as this would streamline exports to North America. Large numbers were built, helping to boost Allied food production in the early 1940s.

Features

The Model N was essentially in the same class as the earlier Model F, though its four-cylinder motor offered a little more horsepower for the three-speed transmission. It had a few minor refinements including better wings and optional pneumatic tyres. The most noticeable change was the overall orange paint scheme. Some wartime production was painted green.

Uses

The Model N carried on from the earlier series as a light ploughing tractor. In wartime it performed yeoman's service at airfields and in many other haulage duties. Collectors enjoy its rugged simplicity and look out for both Irish and English editions.

Related Models

The E27N Major was a modernised, wartime version of similar capacity that evolved into the modern Ford tractor line.

Power & Size

27 hp; weight: 1633kg (3600lb)
length: 259cm (102in)
width: 160cm (63in)
height: 140cm (55in)

Manufacturing & Distribution

The Model N was built in Cork, Ireland, from 1928 to 1932 and then in Dagenham in Essex until 1945. It was a success in the UK, North America, Europe, Australia, New Zealand and southern Africa.

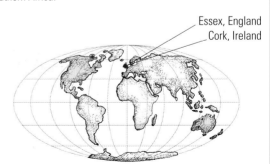

Essex, England
Cork, Ireland

McCORMICK DEERING, MODEL 10-20 GEAR DRIVE

USA, 1936

American inventor Cyrus McCormick was credited with perfecting the reaper, a machine that helped move farming into the mechanical age. In the modern age his name became associated with the International Harvester Company, which brought together the famous McCormick and Deering trade names. The 10-20 Gear Drive gave IHC a modern tractor to compete with Ford.

Features

The successful 10-20 Gear Drive used a four-cylinder IHC motor featuring ball-type main bearings and could burn petrol or kerosene. It had a three-speed transmission and sported a neat, louvred enclosure around the motor. It normally came in dark grey with red wheels, although there was a solid red variation with International badging. The International Junior name was used on early British imports.

Uses

From threshing to ploughing, the 10-20 was an all-purpose tractor. It was advertised as the 'Triple Power Tractor' – for belt, power take-off and drawbar power.

Related Models

The 10-20 was preceded by the more powerful, but superficially similar, 15-30 hp Gear Drive tractor.

Power & Size

10–20 hp; weight: 1678kg (3700lb)
length: 312cm (123in)
width: 152cm (60in)
height: 157cm (62in)

Manufacturing & Distribution

Over 200,000 were built in Chicago, Illinois from 1923 until 1939. Canadian models were often tagged for IHC's Hamilton, Ontario, works. IHC had offices in the UK, and the 10-20 was also sold in mainland Europe, Australia, New Zealand and southern Africa.

Illinois, USA

OLIVER, MODEL 80 STANDARD

USA, 1942

In 1930 the Oliver Farm Equipment Co. became an amalgam of several companies, notably the venerable Hart Parr tractor works of Charles City, Iowa. Despite the Great Depression, Oliver quickly introduced new models that incorporated unit frame construction including the popular MODEL 80 STANDARD. Oliver was a tractor mainstay until it became part of the White Motor Corporation in the 1960s.

Features

The powerful-looking Model 80 used an Oliver/Waukesha four-cylinder motor and a three-speed transmission. It was available on steel wheels or pneumatic tyres with optional electric starters and lights. Oliver favoured utility over style, but the 80 was pleasing to the eye in its Oliver Green hue and red wheels.

Uses

The 80 was rated as a three-plough tractor and accomplished that with ease. A well-placed belt pulley also made it a fine threshing outfit, especially when belted to one of Oliver's famous Red River Special threshing machines. An Oliver 80 can still do a good day's work and then take a break at the local tractor show.

Related Models

A row crop version was also produced. After 1940 a diesel motor was offered for the standard-tread 80. A larger Model 90 was similar in appearance.

Power & Size

40 hp; weight: 1905kg (4200lb)
length: 310cm (122in)
width 155cm (61in)
height: 142cm (56in)

Manufacturing & Distribution

Some 12,000 Model 80s were built in Charles City, Iowa. They were sold under the Cockshutt Plow Co. name in Canada and came to Britain under the Lend-Lease programme early in the Second World War. Some were imported into Australia, New Zealand and southern Africa.

Iowa, USA

DAVID BROWN, MODEL VAK 1

UK, 1941

The David Brown Company had been a producer of machine-cut gears but was a late arrival to the tractor game when it began producing the Ferguson-Brown Model A in 1936. When Ferguson left, David Brown quickly developed its own design, the VAK 1 ('Vehicle Agriculture Kerosene'). It was the beginning of a long, successful run of tractors for the company.

Features

The VAK 1 introduced another colour to the lexicon: Hunting Pink. Its design featured plenty of sheet metal, including a protective shroud ahead of the driver and an almost full covering of the four-cylinder motor. A four-speed transmission provided a good working range and the tractor sported an up-to-date hydraulic lift system, power take-off and belt pulley.

Uses

The VAK 1 was an excellent ploughing tractor capable of pulling three drills with ease. Although a steel-wheeled version was available, on rubber tyres it could attain 18 mph, making it a fine road hauler. The comfortable upholstered seat was wide enough to allow room for a few groceries after a quick trip into town.

Related Models

The VAK 1A was an improved model, followed by the Cropmaster series. The VIG 1/100 was a wartime hauling variant. The VTK 1 was a threshing version.

Power & Size

35 hp; weight: 1474kg (3250lb)
length: 267cm (105in)
width: 169cm (66.5in)
height: 116cm (45.5in)

Manufacturing & Distribution

Some 5350 examples were built from 1939 to 1945 by David Brown at its Meltham works at Huddersfield. The bulk of the sales were in the UK but a few made it to Australia and New Zealand. It helped herald the modern era of farm tractors.

Huddersfield, England

IHC TITAN, MODEL 10-20

USA, 1919

The International Harvester Company began to use the Mogul and Titan trade names just after the turn of the century. One of the last of the breed was the ubiquitous 10-20 TITAN of 1915. This model helped spread the tractor gospel: it heralded an era of less expensive machinery in which the farmer could afford his own small tractor and threshing machine. The Titan was one of the most successful of early lighter designs.

Features

Utilitarian to say the least, the Titan's vital parts were there for all to see. A two-cylinder, horizontal motor could burn kerosene or other low-grade fuels and was mated to a two-speed transmission. It was decked out in basic dark grey with dark red wheels, with the IHC symbol emblazoned on the front of the large water-cooling tank.

Uses

The Titan was an all-purpose tractor and its slow-speed, long-stroke motor made it capable of pulling a three-drill plough with ease. The somewhat primitive look of the Titan makes it an interesting and popular show tractor today.

Related Models

The Titan gave way to the more modern-looking and mechanically improved 10-20 and 15-30 Gear Drive tractors.

Power & Size

10–20 hp; weight: 2495kg (5500lb)
length: 373cm (147in)
width: 152cm (60in)
height 170cm (67in)

Manufacturing & Distribution

Some 78,000 models were built in IHC's factory in Milwaukee, Wisconsin from 1915 until 1922. The Titan helped with wartime ploughing programmes in North America, the UK and France, and was also sold in Australia, New Zealand and southern Africa.

Wisconsin,
USA

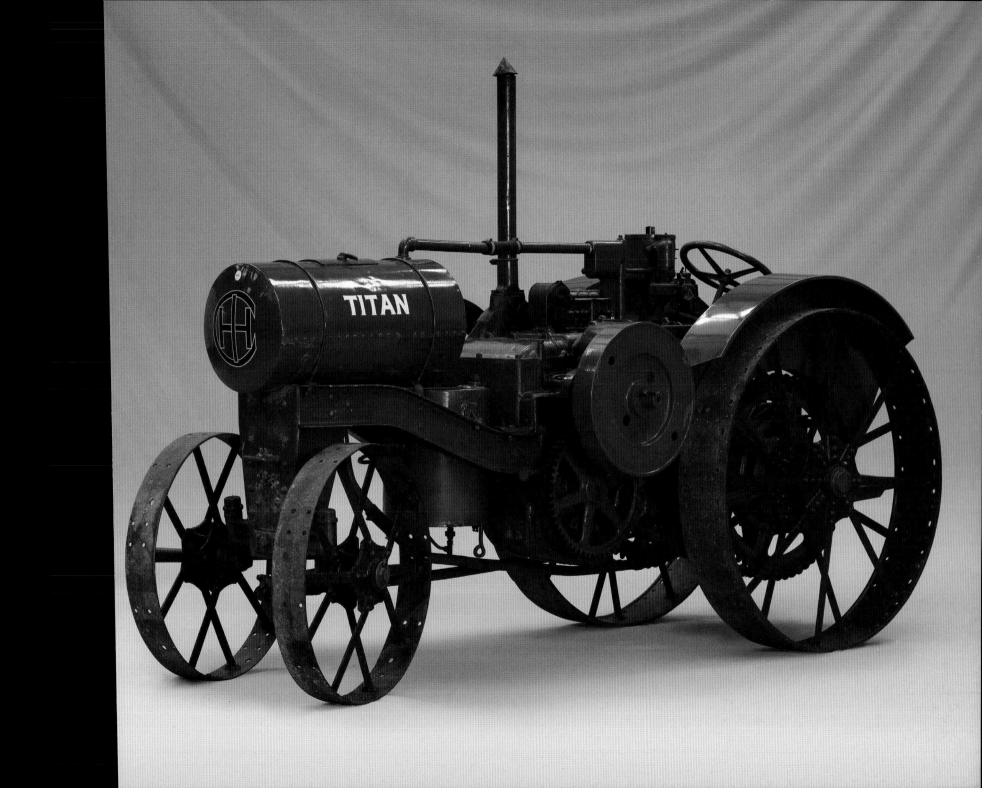

FORD, MODEL 9N

USA, 1939

In 1939 Harry Ferguson joined forces with Henry Ford, who was looking for a modern replacement for his Fordson tractor. The new model was called the FORD 9N with Ferguson System, or more popularly the Ford-Ferguson. The two men fell out after the Second World War: the resulting legal battle over designs and patents was eventually won by Ferguson.

Features

The Ford 9N used a four-cylinder Ford flat head (L head) motor, a three-speed transmission and the Ferguson System hydraulics. The polished aluminium hood and grille accented Ferguson's favorite grey colour on the rest of the tractor. An early model could be bought with steel wheels, and there was a stripped-down wartime version designated the 2N.

Uses

This little outfit was very popular. It filled a niche in the light utility tractor market and was a versatile machine that could perform a wide range of tasks. It can still be found mowing and performing other light duties on farms and estates and is starting to appear at tractor shows and working events.

Related Models

The 9N was replaced in 1947 by the improved model designated the 8N, which remained in production until 1952. The Ferguson TE-20 and TO-20, of similar capacity, competed with the 9N.

Power & Size

23 hp; weight: 1043kg (2300lb)
length: 292cm (115in)
width: 163cm (64in)
height: 132cm (52in)

Manufacturing & Distribution

Some 300,000 of the 9N/2N tractors were built in Dearborn, Michigan from 1939 to 1947. Sold in North America through Ford automotive and Ferguson's own dealers, they were also popular in the UK, Europe, Australia and New Zealand.

Michigan, USA

NUFFIELD UNIVERSAL, MODEL M3

UK, 1949

Britain's economy faced a difficult recovery after the Second World War, so domestic tractor production was encouraged by the government as a way of giving a boost to the economy and aiding the recovery of the farming industry. The Nuffield tractor was named after Lord Nuffield, the name taken by Sir William Morris of Morris Motors. The Nuffield went into production in 1948 with the M3 and M4 UNIVERSAL models.

Features

The M3 tricycle-style or row crop tractor was fitted out with a Morris-Wolsley four-cylinder motor fuelled by distillate. It had five forward speeds and optional hydraulics, three-point hitch, power take-off, belt pulley and lights. The design was solid and utilitarian but spiced up with Nuffield's standard Poppy Orange colour scheme.

Uses

The single front wheel and adjustable rear tread made the M3 an excellent row crop tractor for tilling the fields. The simple construction using sub-assemblies meant easy repair and change-out of essential components.

Related Models

The M4 was the standard front-axle version. The DM3 and PM3 were diesel and petrol variations.

Power & Size

38 hp; weight: 1996kg (4400lb)
length: 312cm (123in)
width: 198cm (78in)
height: 208cm (82in)

Manufacturing & Distribution

The M series was built from 1948 to 1957 in Birmingham. The first production run stayed in Britain to aid post-war reconstruction. M-series tractors were exported to Canada, Europe, Australia, New Zealand, southern Africa and South America.

Birmingham, England

TURNER, YEOMAN OF ENGLAND MARK 2
UK, 1951

Immediately after the Second World War, a great demand for modern farm tractors was hindered by slow deliveries, as the major manufacturers needed time to return to peacetime production. Many new players attempted to fill the gap, including Britain's Turner Manufacturing, which had produced aircraft components during wartime. After some experimentation Turner released the MARK 2, which was advertised with the brand name Yeoman of England.

Features

The novel V-4 diesel motor gave the Mark 2 an interesting look, as the cylinders protruded past the hood on both sides. It had a four-speed transmission and independent brakes. Its appearance was more field-ready than fancy, but it was trimmed with a light-green hood and wings, with red on the wheels and front grille.

Uses

The Mark 2 was a stout machine and could pull three to four ploughs. Some were fitted with a three-point hitch, power take-off and a belt pulley for greater versatility.

Related Models

The Mark 3, with an improved motor, was produced from 1951 until Turner stopped producing tractors in 1955.

Power & Size

40 hp; weight: 2495kg (5500lb)
length: 312cm (123in)
width 203cm (80in)
height: 149cm (58.5in)

Manufacturing & Distribution

The Mark 2 was built at Turner's works in Wolverhampton from 1949 to 1951; around 2500 units were produced. Some were exported to Scandinavia, southern Africa, Australia and New Zealand, but reported mechanical issues and competition from mass-produced tractors affected sales.

Wolverhampton, England

MINNEAPOLIS-MOLINE, MODEL GT

USA, 1941

Three great American tractor companies, Minneapolis Steel & Machinery, Minneapolis Threshing Machine and Moline Plow, joined in 1929 to form Minneapolis-Moline (M-M), which became renowned for its high-quality modern designs, the ultimate being a tractor built with a full cab, radio and heater. The MODEL GT didn't have all of these luxuries but M-M advertised it as 'The Mighty Master of All Jobs'.

Features

The GT was the largest model in M-M's pre-war catalogue and sported a powerful four-cylinder motor and four-speed transmission. M-M's distinctive yellow body colour was named Prairie Gold, and the wheels were painted dark red. The GT was one of a range that M-M described as Visionlined, owing to its clean lines and attractive styling.

Uses

A standard-tread model like the GT was often referred to as a 'wheatland' tractor. Built for heavy drawbar work, the GT was able to pull four to five ploughs. With horsepower to spare, it was also a good threshing tractor.

Related Models

The GT was replaced by the GTA. Later the GB and GVI models used six-cylinder diesel motors and pioneered the use of propane (LPG) fuel.

Power & Size

55 hp; weight: 3084kg (6800lb)
length: 345cm (136in)
width: 183cm (72in)
height: 193cm (76in)

Manufacturing & Distribution

Some 1200 of this model were built in Minneapolis, Minnesota from 1938 to 1941, with strong sales in North America. Before the US entry into the Second World War, many were sent to Britain under the Lend-Lease programme. The GT also enjoyed good sales in Australia and New Zealand.

Minnesota, USA

MASSEY-HARRIS, GENERAL PURPOSE (4WD)

USA, c.1932

The Canadian company Massey-Harris (M-H) built its first tractors in Toronto in 1918. In 1928 M-H purchased the J.I. Case Plow Works of Racine, Wisconsin, which produced the Wallis tractors, and M-H soon centred its tractor production at the old Wallis works. The four-wheel-drive (4WD) GENERAL PURPOSE was the first M-H tractor created by its own designers.

Features

Equipped with a Hercules four-cylinder, 'flat head' engine and three-speed transmission, the General Purpose (GP) was one of the first successful 4WD tractors and was offered in orchard and industrial versions. Styling consisted only of a piece of sheet metal over the motor. Those sold in Britain were dark green with red wheels; in North America they were grey with red wheels.

Uses

The GP was built as a cultivating tractor with adjustable wheel widths, 76-cm (30-in) ground clearance and a 1.8-m (6-ft) turning radius. Some had extensions on the controls so that the operator could drive the tractor whilst sitting on a hay wagon, seed drill or other implement, as he would with a team of horses.

Related Models

An improved GP with overhead valve motor and optional pneumatic tyres was built in 1936 as a last-ditch effort to generate sales.

Power & Size

15–22 hp; weight: 1796kg (3900lb)
length: 310cm (122in)
width: 122–193cm (48–76in, adjustable)
height: 140cm (55in)

Manufacturing & Distribution

Around 3000 were built at Racine, Wisconsin, from 1930 to 1936, and were sold in the UK, France, Canada and the USA. It was particularly popular in Canada for market garden work. The GP heralded the light 4WD tractors that are commonplace today.

Wisconsin, USA

FIELD MARSHALL, MODEL SERIES 3A
UK, 1953

Marshall, Sons & Co. was another of the great old steam builders that tried its hand at internal combustion tractors. However, it hit hard times in the Depression and was taken over by Thomas W. Ward Ltd., and later teamed up with Fowler as Fowler-Marshall Ltd. Post-war tractors such as the SERIES 3A were marketed under the Field Marshall banner.

Features

The Series 3A was the final model using the two-cycle, single-cylinder diesel motor, and had six forward speeds in two ranges, electric start and hydraulic lift. A touch of modern sheet metal could not hide the heavy flywheel and the bulbous stack needed to quiet the thumping single-cylinder power plant. The Series 3A tractor was painted in Fowler's Chrome Orange.

Uses

The single-cylinder motor was no slouch when it came to power, and the Series 3A was advertised as a modern ploughing tractor with plenty of stamina. Its distinctive sound sets it apart from other tractors and still catches people's attention at shows today.

Related Models

The Lanz Bulldog D-4016 was of similar capacity to the Series 3A. Crawler conversions of the Field Marshalls were produced at Fowler's Leeds works. Marshalls went from one cylinder to six in their final series of tractors in the late 1950s.

Power & Size

40 hp; weight: 2722kg (6000lb)
length: 305cm (120in)
width: 193cm (76in)
height: 155cm (61in)

Manufacturing & Distribution

Over 2100 were produced at Marshall's Britannia Works in Gainsborough from 1952 to 1957. The Series 3A was also sold in Canada, South America, Africa, Australia and New Zealand.

Gainsborough,
England

INTERNATIONAL HARVESTER, MODEL 8-16 JUNIOR

USA, c.1919

The unique, lightweight 8-16, with its four-cylinder motor, was an abrupt change from IHC's heavyweight one- and two-cylinder prairie tractors. Simply referred to as the 8-16 or Kerosene Tractor in North America, many were exported to the UK, where they were known as the Junior. The 8-16 JUNIOR came in anticipation of Henry Ford's new lightweight Fordsons.

Features

The 8-16's overhead valve motor and three-speed transmission were borrowed from IHC's truck manufacturing department. A low-slung outfit with plenty of sheet metal, it had a bit of a backwards look as the radiator was placed between the operator and the engine. The grey and red trim would become the company standard for a few years.

Uses

Given its size, the 8-16 was a very capable, light ploughing tractor. It was one of the first tractors produced in large numbers that featured a power take-off, which allowed for a range of operations including hay mowing. This little tractor is now an interesting addition to any show line-up.

Related Models

IHC did not advance the 8-16 design and went on to produce the 10-20 and 15-30 Gear Drive tractors instead.

Power & Size

8-16 hp; weight: 1497kg (3300lb)
length: 335cm (132in)
width: 137cm (54in)
height: 168cm (66in)

Manufacturing & Distribution

Over 33,000 Model 8-16s were reportedly built in Chicago, Illinois from 1918 to 1922. It was a great success and confirmed a demand for light tractors in the USA and Canada. Some 2500 found their way to the UK, many in time to contribute to wartime crop-production efforts.

Illinois, USA

KEROSENE TRACTOR
INTERNATIONAL
8-16
INTERNATIONAL
JUNIOR
KEROSENE TRACTOR

HOLT, MODEL 75

USA, 1918

Benjamin Holt of Stockton, California, was another of the great tractor pioneers. His steam- and petrol-powered crawlers earned the Caterpillar name, trademarked in 1911, for the way they crept across the land. In 1910 Holt opened a second factory in Peoria, Illinois and began manufacture of the successful MODEL 75. In 1925 Holt joined with California rival C.L. Best Co. to create the Caterpillar Tractor Co.

Features

The massive Holt 75 used a four-cylinder petrol motor and had two forward speeds. It was steered via clutches that engaged the appropriate track in combination with the front tiller wheel. The example shown here has been painted military drab, but the standard Holt colour was overall grey. Holt tractors were known for their brawn and not their good looks.

Uses

The Holt 75 was designed for heavy road work and prairie farming. It was pressed into service in the First World War, hauling artillery and supply trains. Heavy and awkward to move, the early crawler tractor is not recommended for the novice collector.

Related Models

The Best 75 from C.L. Best Co. was a competing tractor. The Holt 120 was a grander machine in the same style as the 75.

Power & Size

75 hp; weight: 10,705kg (23,600lb)
length: 610cm (240in)
width: 264cm (104in)
height: 305cm (120in)

Manufacturing & Distribution

Produced in Stockton, California from 1913 to 1921 and Peoria, Illinois until 1924, the Holt 75 was successful throughout North America and exported to the UK, France and Russia. Ruston, Proctor & Co. of Lincolnshire built them in England under licence for the British Ministry of Munitions in 1916 and 1917.

Illinois, USA

Lincolnshire, England

California, USA

MASSEY-FERGUSON, MODEL 35X

UK, 1963

In 1953 Massey-Harris purchased Harry Ferguson's tractor business. The company was known as Massey-Harris-Ferguson until 1957, when it became Massey-Ferguson (M-F). The popular Model 35 tractor began life under Ferguson and endured under M-F, built in both the UK and the USA. The model went through several iterations, culminating in the MODEL 35X.

Features

The 35X sported the remarkable three-cylinder Perkins diesel motor and could be purchased as the Deluxe model, with six speeds in dual range, or the Multi Power, with Massey's novel hydraulic 'on-the-go' clutch system, which offered 12 speeds. The 35X was amongst the last with the rounded-front Ferguson styling. The drive train and wheels retained the traditional Ferguson Grey, with the sheet metal in red.

Uses

The 35X was a truly modern tractor with diesel power, Ferguson hydraulics and great handling abilities. The post-war styling makes the 35X a desirable collector's item and it is presented in tractor shows as a latter-day classic.

Related Models

The restyled M-F 135 continued the 35 tradition and led to the Model 245. The 65 was a more powerful tractor in the same style.

Power & Size

44.5 hp; weight: 1451kg (3200lb)
length: 297cm (117in)
width: 163cm (64in)
height: 137cm (54in)

Manufacturing & Distribution

The 35X was built in Coventry until 1964. It was especially popular in southern Africa and also saw service in Canada, Australia, New Zealand and mainland Europe. The 35X was the last of the many variants of the highly successful 35 series.

Coventry, England

FORDSON, MODEL E1A-DKN

UK, 1952

Fordson had wanted to build the E1A earlier, to replace the F and N series, but was thwarted by wartime exigencies. When it finally appeared the new E1A and its DKN variant was referred to as the New Major; a modern diesel motor and improved hydraulics were its chief selling points. The Fordson name remained in use for UK-built tractors until the early 1960s.

Features

Rather than the normal diesel power plant, the E1A-DKN was fitted with Ford's new overhead valve four-cylinder motor, capable of burning petrol or lower-grade fuels. It sported a dual range transmission with six forward speeds. A more streamlined look set it apart from the previous E27N Major and before long the lighter blue colour became the company standard. The orange wheels would soon give way to white.

Uses

The E1A-DKN was a well-built machine that could handle almost any job one could imagine. It was useful in areas where diesel fuel was harder to obtain. A somewhat scarce model today, it is an especially interesting machine for Fordson enthusiasts.

Related Models

Later versions were the improved Super Major, Power Major and Dexta. The E1A chassis often formed the basis for after-market 4WD conversions.

Power & Size

35 hp; weight: 2404kg (5300lb)
length: 333cm (131in)
width: 165cm (65in)
height: 160cm (63in)

Manufacturing & Distribution

Well over 200,000 of the E1A series were built in Dagenham in Essex from 1951 to 1958. Large numbers were exported to North and South America, Australia, New Zealand, Europe and southern Africa.

Essex, England

ADVANCE RUMELY, OIL PULL, MODEL 20-35 M

USA, 1925

Advance Rumely's 20-35 M was one of the company's Lightweight series, scaled down from the earlier prairie tractors. Rumely perfected the burning of lower-grade fuels such as kerosene and distillates through a patented carburetion system. The OIL PULL name was conceived as a catchy trademark that was soon recognised throughout the world.

Features

The two-cylinder horizontal motor of the 20-35 M used light oil for cooling, allowing for higher engine temperatures and the cleaner burning of kerosene. The engine exhaust was blasted through a nozzle atop the distinctive radiator, inducing a draught of cooling air through it. The 20-35 M operated with a three-speed transmission. Its styling was utilitarian at best, but it was tastefully decked out in Brewster Green, with red pinstripes on the wheel spokes.

Uses

The 20-35 M was marketed as a middle-range threshing tractor, although it was adept at drawbar work. With a unique look and sound, it is the gold standard for today's tractor collectors.

Related Models

The 15-25 L, 25-45 R and 30-60 S were companion models in the Lightweight series. They were all replaced by the Super-Powered Lightweight series of 1928.

Power & Size

20–35 hp; weight: 3946kg (8700lb)
length: 381cm (150in)
width: 183cm (72in)
height: 254cm (100in)

Manufacturing & Distribution

Over 3600 were built by Advance Rumely Thresher Co. at La Porte, Indiana from 1924 to 1927. Rumely had a well-developed dealer network throughout the USA and Canada, and in South America.

Indiana, USA

NUFFIELD, MODEL 3/42

UK, 1961

Produced by the British Motor Corporation (BMC), Nuffield tractors marketed under the Universal name gained a strong following after the Second World War. The MODEL 3/42 was introduced in 1961, with a modern diesel motor and updated hydraulic system. In 1968 British Leyland took over from BMC and the Nuffield name joined the tractor industry's long list of fallen flags.

Features

The 3/42 moniker stood for the three-cylinder diesel motor with 42 hp output. It used a five-speed transmission, modern hydraulics and depth control. The 3/42 retained Nuffield's pleasant 1950s-style rounded hood and sported an overall treatment of Poppy Orange, soon to be replaced by Leyland blue.

Uses

The versatile and well-built 3/42 was a modern tractor in every way. It was a good second tractor on a large farm or could handle the full workload of a smaller operation. The inexpensive canopy softened the morning chill of autumn ploughing. At today's shows and fairs a Nuffield will always be cheerfully appreciated in the 'classic' category.

Related Models

The 4/60 was a larger four-cylinder companion to the 3/42. Both were replaced by the 10/42 and 10/60 respectively.

Power & Size

42 hp; weight: 2404kg (5300lb)
length: 305cm (120in)
width: 183cm (72in)
height: 193cm (76in)

Manufacturing & Distribution

Production began at BMC's Birmingham works, but was shifted near Edinburgh during 1961. The series was built from 1961 to 1964 and sold in the UK, Europe, Canada, Australia, New Zealand, southern Africa and South America.

Birmingham, England
Edinburgh, Scotland

FOWLER, MODEL CHALLENGER III

UK, 1951

The John Fowler Co. was a world pioneer in steam ploughing and threshing equipment, shifting to internal combustion crawler tractors as the steam age waned. The company developed a line of modern diesel crawlers during the Second World War, but in 1947 it was taken over by Thomas W. Ward Ltd., under whose auspices the CHALLENGER III was introduced.

Features

The Challenger series appeared in 1950. The Challenger III was the 'light heavyweight' model and could be purchased with either a Leyland or a Meadows six-cylinder diesel motor; both were British-built and of comparable power. The Challenger III operated in six forward speeds. The series was utilitarian in appearance and painted entirely in Fowler's Chrome Orange.

Uses

The Challenger III crawler saw extensive military service but was also at home in heavy field work. A number of dozer-blade attachments were available for construction and road building. Custom units served the Royal National Lifeboat Institution (RNLI) through the 1950s.

Related Models

The Challenger III was accompanied by the Challenger I, II and IV at 50, 80 and 150 hp respectively, and succeeded by the heavier and more powerful Mark 33.

Power & Size

95 hp; weight: 11,113kg (24,500lb)
length: 401cm (158in)
width: 234cm (92in)
height: 198cm (78in)

Manufacturing & Distribution

Under Thomas W. Ward Ltd., the Challenger III was built at Fowler's works in Leeds from 1950 to 1956, and sold in the UK, Europe, Australia, New Zealand, Asia and southern Africa.

Leeds, England

HART PARR, MODEL 18-36 G

USA, 1927

Hart Parr is believed to be the first company to have used the term 'tractor'. In those days it was best known for its heavyweight prairie tractors, but the new Hart Parr of 1918 was the first in a long line of general-purpose machines. Amongst them was the 18-36 G, a mid-sized model that would see Hart Parr through to 1929, when it became part of Oliver.

Features

The 18-36 G was powered by a two-cylinder horizontal motor matched with a two-speed transmission. It used traditional construction techniques with a heavy channel, steel frame supporting the motor and drive components, but was nonetheless a good-looking and well-designed machine. Hart Parr's dark green paint scheme with red wheels would last into the Oliver years.

Uses

Although perfectly acceptable for field work, the 18-36 was intended to compete with the likes of the Rumely Oil Pull as a threshing tractor. It took some getting used to, as the Hart Parr's drive pulley was on the opposite side of the tractor to most and was somewhat obscured by the sheet metal.

Related Models

The 18-36 G was was replaced by the 18-36 H, with three forward speeds rather than two. The Hart Parr 12-24 and 28-50 had different outputs but were of the same design as the 18-36 G.

Power & Size

18–36 hp; weight: 2812kg (6200lb)
length: 335cm (132in)
width: 185cm (73in)
height: 155cm (61in)

Manufacturing & Distribution

Built in Charles City, Iowa, from 1926 to 1930 but not continued under Oliver. Very popular in the USA and Canada, the 18-36 G also sold in the UK, mainland Europe, Australia and New Zealand.

Iowa, USA

WATERLOO BOY, MODEL N

USA, 1920

The Waterloo Gasoline Engine Co. of Waterloo, Iowa, was purchased by John Deere in 1918. Waterloo's tractor pedigree went back to John Froehlich's experiments of 1892, although the Waterloo Boy series didn't appear until 1912. The MODEL N was the first tractor tested in the famous University of Nebraska Tests and formed the basis of John Deere's successful line that endures to this day.

Features

The Model N had a two-cylinder horizontal motor and two forward speeds. The old-fashioned channel-frame construction included chain-type steering, although this was later changed to automotive-type steering (often called Ackerman steering). The standard colour was green with yellow wheels, although a variation added a dark red to just the motor.

Uses

The Model N was a good general-purpose machine and did yeoman's service in the UK and North America, ploughing the land for the First World War. The belt pulley was well positioned for trouble-free connections to the threshing machine.

Related Models

The single-speed Model R preceded the Model N. The successful John Deere-designed Model D continued the two-cylinder tradition of the Waterloo Boy tractors.

Power & Size

12–25 hp; weight: 2767kg (6100lb)
length: 335cm (132in)
width: 183cm (72in)
height: 160cm (63in)

Manufacturing & Distribution

Over 21,000 were built from 1917 to 1924 in Waterloo, Iowa, and distributed across North America. It was sold in the UK by the Overtime Tractor Co. under the Overtime brand name. The Model N was also sold across Europe, and in southern Africa, Australia and New Zealand.

Iowa, USA

LANZ BULLDOG, MODEL D7506

GERMANY, c.1935

The MODEL D7506 was produced by one of Europe's most successful tractor builders, Heinrich Lanz AG. Lanz adopted the brand name Bulldog to stress the tough and tenacious qualities of their machines. The Lanz design endured from its inception in 1921 until the company was taken over by John Deere in 1956.

Features

The single-cylinder motor was of a two-cycle design and used hot-bulb ignition, which meant preheating the combustion chamber with a blowlamp before firing up the motor. After that, it would burn any fuel down to crude or waste motor oil. The D7506 transmission offered six forward speeds in two ranges. The colour scheme was basic grey with red wheels until John Deere Green took over.

Uses

The D7506 was a multipurpose outfit: strong on the drawbar, ready to thresh at a moment's notice, and a capable road hauler. Rugged simplicity was the Lanz hallmark and the tractor gave countless hours of trouble-free service. The booming cadence of the D7506's single-cylinder motor is a real crowd-pleaser.

Related Models

Several companion sizes of up to 55 hp were built, along with improved post-war models. The British Field Marshall tractors followed the Lanz design.

Power & Size

25 hp; weight: 2313kg (5100lb)
length: 279cm (110in)
width: 160cm (63in)
height: 178cm (70in)

Manufacturing & Distribution

The D7506 Bulldog was built in Mannheim, Germany, from 1935 to 1952. Lanz tractors were sold in the UK, Europe, Canada, Australia and New Zealand and were built locally in Argentina, France, Spain and Poland.

Mannheim, Germany

SAUNDERSON, MODEL G UNIVERSAL

UK, 1916

Tractor pioneer Herbert Saunderson, who came from Bedfordshire, was building lightweight tractors under the Universal name when most everyone else was turning out giants. They ranged from his dandy 6-8 hp Little Knock-About up to the 45 hp Colonial. Around 1915 he produced three new Universal models including the popular, mid-sized MODEL G.

Features

The early Saundersons had mid-ships or forward operator positions whilst the Model G employed a conventional seat placement. It sported a two-cylinder vertical motor, three forward speeds, a stout channel frame and heavy cast wheels. A later version even showed a glint of modernistic styling. The pictured example is in standard Saunderson green trim with red wheels.

Uses

The Model G came out just in time for the critical wartime ploughing effort. Clear sightlines made it an easy tractor to belt up to the threshing machine. A number of Universals have survived for the enjoyment of rally-goers today.

Related Models

Saunderson's Models J and B filled out the 1915 series. The Light and Super Light-Weight models closed out Saunderson's production after the war.

Power & Size

23–25 hp; weight: 1860kg (4100lb)
length: 366cm (144in)
width: 168cm (66in)
height: 213cm (84in)

Manufacturing & Distribution

Built at Saunderson's Elstow Works in Bedfordshire, the machine was also exported to Australia, New Zealand and Africa. Saunderson also operated British Canadian Agricultural Tractors Ltd. in Saskatoon, Canada.

Bedfordshire, England

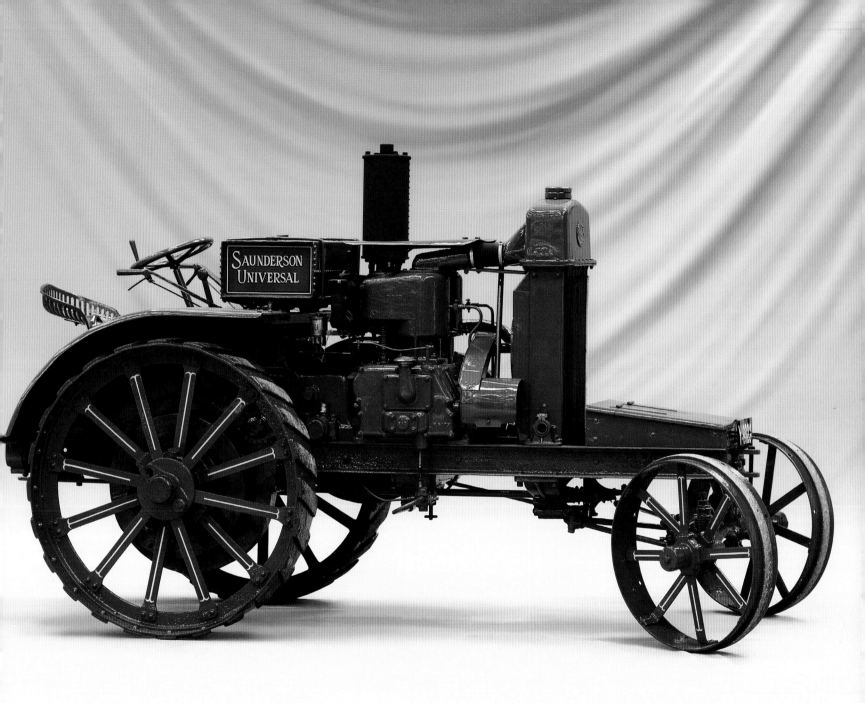

WEEKS-DUNGEY, MODEL NEW SIMPLEX

UK, c.1922

The history books do not give his first name, but a Mr Dungey from Kent needed a farm tractor. He talked a local iron foundry and engine-making firm, William Weeks & Son, into making one for him. Called the Simplex, it was built in 1915 and, although rudimentary in appearance, it worked well enough to encourage Weeks to carry on. His more substantial New Simplex made its first appearance in 1919.

Features

Weeks was lucky enough to have secured some American-made, four-cylinder Waukesha motors and three-speed transmissions. Differential lock and rack-and-pinion steering were an advance over the competition. The New Simplex had solid wheels, a heavy frame and a hint of styling in the grey tin work and red wheels. A rubber tyre version was available for road hauling.

Uses

Mr Dungey had a large orchard and hop-growing operation and needed a tractor that wouldn't damage the tender crops. This led Weeks to give the New Simplex a compact layout.

Related Models

Sadly, although orchard tractors were eventually offered by most of the major tractor builders, this model marked the end of Weeks's tractor production.

Power & Size

30 hp; weight: 1497kg (3300lb)
length: 249cm (98in)
width: 122cm (48in)
height: 137cm (54in)

Manufacturing & Distribution

Built at the Perseverance Iron Works in Maidstone, Kent from 1919 to c.1925. It is thought that only around 220 examples were produced. The tractor seems to have been limited mainly to local sales.

Kent, England

INTERNATIONAL HARVESTER, MODEL W4

USA, 1940

After the First World War, IHC became one of the more dynamic tractor builders, with an extensive range of rubber tyre and crawler tractors. Along with wide-tread models like the W4, they offered a full range of row crop machines under the new Farmall brand. The company endured until 1986, when it joined with Case as part of the Tenneco corporate empire.

Features

The W4's four-cylinder motor could burn petrol and distillates and was mated to a four-speed transmission. The tractor was available on steel wheels but pneumatic tyres had gained popularity. IHC was amongst the manufacturers who employed the American industrial designer Raymond Loewy to give the W4 and sister models a modernistic look. International Red paint replaced their traditional grey in around 1937.

Uses

The W4 was the modern workhorse, a sturdy, standard-tread tractor for drawbar work. With a side-mounted belt pulley for threshing and a power take-off at the rear for mowing or baling, the W4 was a master of all trades.

Related Models

This model was accompanied on dealers' lots by the larger W6 and W9 and was replaced by the five-speed Super W4 in 1953.

Power & Size

22–24 hp (distillate-petrol)
weight: 1905kg (4200lb)
length: 287cm (113in)
width: 147cm (58in)
height: 140cm (55in)

Manufacturing & Distribution

Well over 24,000 W4s were built in Chicago, Illinois, from 1940 to 1953, and were sold in North America, the UK, Europe, Australia and New Zealand. After the war, IHC opened a factory in Doncaster, England.

Illinois, USA

DAVID BROWN, MODEL 1210

UK, 1976

After collaborating with Harry Ferguson in the 1930s, David Brown & Sons charted its own course with a full range of first-class tractor designs. The 1210 was part of a modern line of tractors introduced just before the company was taken over by the American corporate giant Tenneco in 1972. The David Brown name was dropped in around 1983.

Features

David Brown built the four-cylinder diesel motor and the excellent 12-speed syncromesh transmission used in the 1210. This model used up-to-date hydraulics and live power take-off and, as pictured, could be made into a four-wheel drive machine by fitting the optional front drive axle. After 1965 David Brown tractors were painted with Orchid White sheet metal and Poppy Red castings and wheels.

Uses

The 4WD option and the multispeed 'tranny' made this a truly general-purpose tractor ready to work hard in all field conditions. With the company now defunct, David Brown tractors are becoming popular with collectors.

Related Models

The companion 1212 model used an automatic Hydrashift transmission. The 1410 and 1412 each had a higher hp.

Power & Size

72 hp; weight: 3447kg (7600lb)
length: 143in (363cm)
width: 142–203cm (56–80 in, adjustable)
height: 272cm (107in)

Manufacturing & Distribution

Some components were made at David Brown's Leigh works near Manchester with final assembly at Meltham in Huddersfield. Production of the 1210 ran from 1971 until 1980. Extensive exporting saw the tractors reach North and South America, Europe, southern Africa, Australia and New Zealand.

Huddersfield, England

INTERNATIONAL HARVESTER, MODEL T-20 CRAWLER

USA, 1933

International Harvester built a strong following for its T ('tracked') series crawler tractors (TD for diesels). Its first effort was the 1929 TracTracTor, a crawler conversion of the 10-20 Gear Drive wheeled tractor. The T-20 was the first true IHC crawler designed as such and continued to use the catchy TracTracTor brand name.

Related Models

The T-35 and T-40 were larger units in the series. The T-20 was replaced in 1940 by the T-6 and TD-6.

Power & Size

25 hp; weight: 3175kg (7000lb)
length: 312cm (123in)
width: 140cm (55in)
height: 142cm (56in)

Features

The T-20 used IHC's own four-cylinder petrol and kerosene motor and a three-speed transmission, and could be purchased with power take-off, belt pulley and lights. The louvred engine panels gave it a modern look but it retained IHC's standard grey colour scheme until the company switched to red around 1937. The T-20's compact layout made it easy to operate.

Manufacturing & Distribution

Over 15,000 were built from 1931 to 1939 at IHC's Chicago, Illinois, tractor works. This model saw widespread distribution through IHC dealers across North America, as well as in the UK, Europe, South America, Australia and New Zealand.

Uses

The T-20 was meant mainly as a farm tractor but could be used in construction and in the timberlands. The T-20 could do any of the wheeled tractor's farm jobs but offered rugged crawler capabilities for tough situations. There was also an orchard version with extra shielding over the tracks to protect tender plants and branches.

Illinois, USA

CASE, MODEL CC-3

USA, 1936

Old Abe was the name of the Case eagle mascot, and he watched over many years of solid success with steam traction engines and heavy prairie tractors. By the 1930s Case was ready to try its hand at lighter, row crop tractors such as the CC-3, which led to a long line of popular cultivating tractors. Today the Case name endures as part of the worldwide Case New Holland farm equipment empire.

Features

The innovative CC-3 featured a four-cylinder motor capable of burning kerosene, and had three forward gears. The CC-3 row crop or tricycle model was accompanied by a standard-tread CC-4, but farmers could purchase a tractor with both front ends and change from standard to row crop. These were amongst the last Case tractors with the grey chassis and red wheels.

Uses

The CC-3, with adjustable wheel widths and an assortment of mountable tillage equipment, was ideal for corn cultivation. The tricycle layout and vintage styling make it a sought-after collector's item today.

Related Models

The CC-3 was succeeded by the RC and later the VC and VAC. The CC-4 was the standard tread version to the CC-3.

Power & Size

27 hp; weight: 1860kg (4100lb)
length: 348cm (137in)
width: 122–213cm (48–84in, adjustable)
height: 145cm (57in)

Manufacturing & Distribution

The CC-3 was built by the J.I. Case Threshing Machine Co. in Racine, Wisconsin, from 1929 to 1939 and was very popular in the Midwest Corn Belt of North America. It also enjoyed sales in the UK, Europe, Australia and New Zealand.

Wisconsin,
USA

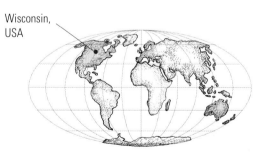

BRITISH WALLIS

UK, c.1925

Two separate companies used the Case name, both based in Racine, Wisconsin. One, the J.I. Case Threshing Machine Co., evolved into the name seen on tractors today; the other, the J.I. Case Plow Works, built the Wallis tractors, named for company president Henry Wallis. In 1919 Ruston & Hornsby were licensed to build the Wallis OK model in the UK, calling it the BRITISH WALLIS.

Features

The British Wallis used a locally produced Ruston four-cylinder motor, burning petrol and kerosene. The tractor had two forward speeds. Wallis was known for its unit frame design incorporating a rolled U-shaped plate to form the main frame. This unit frame had easily removed access ports that allowed the owner to readily service the motor and transmission. The British Wallis had a modernistic design highlighted in a pleasant green hue with red wheels.

Uses

The British Wallis was a fine general-purpose tractor capable of a wide range of field duties along with threshing. This model is now a very rare and desirable tractor for collectors.

Related Models

The British Wallis was discontinued but Massey-Harris, who bought J.I. Case Plow Works in 1928, continued the Wallis design until the Second World War with its Pacemaker, Challenger and Model 25.

Power & Size

28 hp; weight: 1678kg (3700lb)
length: 335cm (132in)
width: 155cm (61in)
height: 160cm (63in)

Manufacturing & Distribution

The British Wallis was built from 1919 until *c.*1928 in Lincoln. These tractors also appear in Australia and New Zealand.

Lincoln, England

CHAMBERLAIN, MODEL SUPER 70
AUSTRALIA, 1960

Chamberlain Industries moved into an old armaments factory at Welshpool, Australia, after the Second World War and joined an impressive list of Australian tractor producers. Their first example, the 40K, came out in 1949. The later SUPER 70 was advertised as the 'Powerhouse On Wheels' and its high construction quality allowed it to compete well with the larger companies exporting to Australia.

Features

For the Super 70, Chamberlain went from its own two-cylinder motors to the powerful General Motors (Detroit) 3-71 two-stroke diesel, mated to a nine-speed transmission. The 'Screaming Jimmy' motor, as it was often called, was music to the ears under a full load. The Super 70 had a no-frills look with its overall orange colour scheme.

Uses

The Super 70 could be counted on to put in many long hard days working the fields. Chamberlain made implements such as ploughs to go with the tractor and also supplied a pulley attachment for powering belt-driven machinery.

Related Models

The Super 90 was a more powerful addition to the Chamberlain catalogue. The company's later Champion series tractors had six-speed transmissions.

Power & Size

70 hp; weight: 4400kg (9700lb)
length: 356cm (140in)
width: 178cm (70in)
height: 224cm (88in)

Manufacturing & Distribution

The Super 70 was built at Welshpool, near Perth in Western Australia, from 1955 to 1962. Sales were strongest in Australia and New Zealand, with some units reaching southern Africa and the UK.

Perth, Australia

MASSEY-FERGUSON, MODEL 135
UK, 1969

Massey-Ferguson's MODEL 135 was one of the smallest of the 100 Series or 'Red Giant' line that would eventually include models up to 120 hp. Based in Toronto, Canada, Massey-Ferguson had manufacturing facilities in several countries. Changing financial fortunes led it to become part of the AGCO farm equipment empire in 1994, though the M-F brand is still in use.

Features

The Perkins three-cylinder diesel was the engine of choice for the 135, with the option of a four-cylinder petrol motor. The standard transmission offered six speeds in two ranges, but Multipower was also available. The 135 had dark red wings and hood with the grille trimmed in Silver Mist, a colour often used on the wheels as an alternative to Ferguson Grey.

Uses

The 135 was the quintessential general-purpose tractor and thousands of them still perform daily duties. The Ferguson System hitch meant that a wide variety of rear-mounted implements could be used. Collectors are now restoring the ubiquitous 135s that they admired in their childhood.

Related Models

The 135 replaced the old 35 series and was superseded by the M-F 245, a restyled mid-1970s tractor.

Power & Size

45 hp; weight: 1451kg (3200lb)
length: 300cm (118in)
width: 163cm (64in)
height: 150cm (59in)

Manufacturing & Distribution

Over 350,000 were built at Coventry from 1965 until the mid 1970s. Many were also built in Detroit, Michigan. The 135 was distributed in the UK, the Americas, Europe, Africa, India, Australia and New Zealand.

Michigan, USA

Coventry, England

REPORTAGE

If a man's home is his CASTLE, then his tractor is both his *indispensable squire* and trusted STEED. Keeping that armour in good condition is a challenge. Restoring and maintaining tractors is a *labour of love,* and PLOUGHING matches at working events offer owners an opportunity to test both METAL and *mettle.*

Maintaining the Paul Rackham collection, UK

Fix the engine and then shine up the Motorway Yellow livery till I can see my face in it!

Harry Ferguson, Inc.
DETROIT, MICHIGAN
MANUFACTURED IN U.S.A. BY
CITY M'F'G CO.
INCORPORATED
CINE, WISCONSIN
SERIAL

INTERNATIONAL
JUNIOR
KEROSENE TRACTOR

To do list
Keep all vintage
tractors working
like clockwork...

Remember to check the overhead
valve four-cylinder unit.

JOHN DEERE

LANZ
22 PS
BULLDOG

The
Weald of Kent
Ploughing Match,
UK

Tractor owners leave
no land unturned.

Motto in life
and on a tractor:
keep on the
straight and
narrow.

It's important to plough your
own six or eight drills in life.

Opening: well cut, uniform, straight
Start: uniform
Seed bed: weed control and soil made available
Firmness: firm, well-packed drills
Uniformity: clearly defined, straight drills, no pairing
Finish: uniform, shallow and straight
Ins & Outs: neat and regular
General appearance: vintage charm!

Boy, this is fun!

staying cool under
pressure is key.

I'm the Corduroy king
- master of straight
parallel lines.

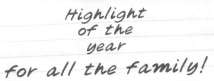

Four-legged fun on a three-drill day.

Highlight
of the
year
for all the family!

You are never too young OR too old to compete!

New and vintage, big daddies and small versions, humans and tractors love a ploughing match and a prize!

Holiday Photos
from
New York State &
Vermont, USA

My holiday photos are a little different from most...It's Iowa for me next year.

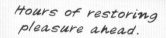

Hours of restoring pleasure ahead.

Couldn't help snapping this one up!

Wish list:

Save up for a Johnny Popper next year!

They don't call 'em Deere for nothing!

110

GLOSSARY

Auxiliary motor Small petrol motor used for starting the larger, main diesel motor in a tractor. Sometimes called a pup, pony or donkey motor.

Belt pulley Powered attachment for driving a flat belt to operate a threshing machine or other implement.

Depth (draft) control A system – such as the Ferguson System – that automatically adjusts implement depth and draft using hydraulic pressure.

Differential lock A device to ensure that all drive wheels are engaged.

Displacement Standard measure of the volume displaced by one stroke of the piston in a cylinder, expressed in cubic centimetres or cubic inches.

Distillate Lower-grade fuel also called 'tractor vaporising oil' (TVO) in the UK.

Dual range Shifting gears that offer high and low range, doubling the usual transmission gear selection.

Flat head motor Motor with the intake and exhaust valves usually in a position parallel to the cylinders. Also called L head motor.

Four-cycle motor The most common tractor motor, which has a sequence of piston travel involving intake, compression, ignition (expansion) and exhaust strokes.

Heavyweight tractor Early tractors with channel steel frame members and minimal styling.

Horsepower (hp) Standard measure of power output. Can be stated as engine, drawbar or brake horsepower. Early advertised ratings expressed drawbar-belt pulley horsepowers, i.e., 10-20, 8-16.

Hot bulb A device that uses a flame to preheat the motor's intake to enable ignition of lower grade fuels.

Hydraulics Internal system pumping light oil for raising and lowering implements, or for running remote implement drives.

Independent brakes Separate foot or hand controls for the right and left brakes.

Lend Lease A programme whereby the USA provided equipment to the UK and her Allies prior to the US entry into the Second World War.

Lightweight tractor A later style of tractor using unit frames.

Overhead valves (OHV) A motor design in which the intake and exhaust valves are above the pistons.

Power take-off (pto) Normally a splined shaft driven through the transmission onto which a longer drive shaft could be attached to power an implement such as a mower.

Row crop A tractor with higher ground clearance and adjustable wheel widths for cultivating taller crops such as corn. Generally tricycle or narrow front style but some standard-tread tractors also fall into this category.

Standard tread A tractor with front and rear wheels set at approximately the same widths.

Three-point hitch (3ph) Rear-mounted lifting system for attaching implements to a tractor.

Torque tube A style of frame with a long tube between the motor and final drives.

Two-cycle motor A sequence of piston travel involving a compression stroke followed by a combined ignition, expansion, exhaust and intake stroke.

Unit frame The motor, transmission and final drive castings also serve as the tractor frame.

Wheatland tractor Standard-tread tractor with wide rubber tyres used mainly for drawbar work.

PICTURE CREDITS
Istockphoto/Ralf Hettler: p10
SF photo/Shutterstock.com p12
Istockphoto/Roman Ponomarets: p13
Istockphoto/SimplyCreativePhotography: p14

SHOWS

BRITAIN

Carrington Steam & Vintage Tractor Rally
Last weekend of May.
White House Farm, Main Road, Carrington
PE22 7DZ, UK
Telephone: +44 (0)7702 208518
Website: www.carringtonrally.co.uk
Email: enquiries@carrington.co.uk

Great Dorset Steam Fair
Early September.
Dairy House Farm, Child Okeford, Blandford, Dorset DT11 8HT, UK
Telephone: +44 (0)1258 860361
Website: www.gdsf.co.uk
Email: enquiries@gdsf.co.uk

Little Casterton Working Weekend
Mid September.
Stamford, Lincolnshire, UK
Website: www.fensvintage.co.uk

Newby Hall Rally
Mid June.
Yorkshire Vintage Association
73 Church Road, Wawne, Hull HU7 5XL, UK
Telephone: +44 (0)7711 692378
Website: www.theyva.com

AUSTRALIA

Lake Goldsmith Steam Rally
Check website for show dates.
PO Box 21, Beaufort, Victoria 3373, Australia
Telephone: +61 35349 5512
Website: www.lakegoldsmithsteamrally.org.au
Email: info@lakegoldsmithsteamrally.org.au

CANADA

Harvest Festival
Early September.
Reynolds-Alberta Museum
Box 6360, Wetaskiwin, Alberta, CA, T9A 2G1
Telephone: 1-780-361-1351
Website: www.machinemuseum.net
Email: reynoldsalbertamuseum@gov.ab.ca

Heritage Farm Show
Mid August.
Bruce County Heritage Association
Box 128, Paisley, Ontario, CA, N0G 2N0
Telephone: 1-519-389-8288
Website: www.bruceheritage.org

Thresherman's Reunion & Stampede
Late July.

Manitoba Agricultural Museum
Box 10, Austin, Manitoba, CA, R0H 0C0
Telephone: 1-204-637-2354
Website: www.ag-museum.mb.ca
Email: info@ag-museum.com

Western Development Museums, Saskatchewan
Thresherman's Show, Yorkton, late July; Pion-Era, Saskatoon, early July; Those Were The Days, North Battleford, early August.
2610 Lorne Avenue South, Saskatoon
Saskatchewan, CA, S7J 0S6
Telephone: 1-306-931-1910
Website: www.wdm.ca
Email: saskatoon@wdm.ca

EUROPE

Die Schlepperfreunde im Rheinland
Every two years. Check website for dates.
Kühlerhof Farm, Hückelhoven-Doveren
Lanz-Bulldog, Verein "West" e.V
Vorstadt 23, D-41812 Erkelenz, Germany
Telephone: +49 (0)2435 2380
Website: www.lanzbulldog.de

Historischer Feldtag
Mid August.
Treckerveteranenclub Nordhorn
Bahnhofstr. 21, 48529 Nordhorn, Germany
Telephone: +49 01578 7638852
Website: www.treckerclub.de
Email: info@treckerclub.de

International Historisch Festival
Late July.
Historische Motoren en Tractoren Vereniging
Ruiterrein De Vossberg, Ninnesweg 176
5981 PD, Panningen, The Netherlands
Website: www.hmtklep.nl

La Musée de la Machine Agricole Ancienne
Mid August.
9 Rue du Maître de Forges, Le Bourg
58200, Saint-Loup (Burgundy), France
Telephone: 00 33 3 86 39 91 41
Website: www.framaa.fr

Vintage Tractor Show
Mid August.
Foundation Beleef Historische Landbouw
Kapelweg 19a, 5571 XD, Bergeijk
The Netherlands
Website: www.tractorshowbergeijk.nl

NEW ZEALAND
New Zealand Vintage Machinery Club, Canterbury
Check website for dates and locations.
PO Box 20082, Christchurch 8030, NZ
Website: www.nzvintageclub.co.nz
Email: nzvintagenews@xtra.co.nz

SOUTH AFRICA
Veteran Tractor Exhibition
Early September.
South African Federation of Vintage Tractors
& Engines Clubs
Website: www.savtec.co.za

UNITED STATES
Divide County Threshing Bee
Mid July.
Crosby, North Dakota, USA
Website: www.dcthreshingbee.com

Eastern Shores Threshermen & Collectors Association
Early August.
5946 Federalsburg Road
Federalsburg, Maryland 21632, USA
Telephone: 1-410-673-2414
Website: www.threshermen.org
Email: threshermen@toad.net

Puget Sound Antique Tractor & Machinery Association
Early August.
Box 255
Lynden, Washington 98264, USA
Website: www.psatma.com
Email: psatandma@gmail.com

Steam, Gas & Threshing Show
Mid August.
American Thresherman Association
627 Ballpark Road
Pinckneyville, Illinois 62274, USA
Telephone: 1-618-357-6643
Website: www.americanthresherman.com

Western Minnesota Steam Threshers Reunion at Rollag
Early September.
Box 9337
Fargo, North Dakota 58106, USA
Telephone: 1-701-212-2034
Website: www.rollag.com
Email: secretary@rollag.com

ASSOCIATIONS

ALLIS-CHALMERS
Old Allis News
471 70th Avenue, Clayton, Wisconsin 54004, USA
Telephone: 1-715-268-4632
Email: oldallisnews@amerytel.net

J. I. CASE
J. I. Case Heritage Foundation
Box 081156, Racine, Wisconsin 53408, USA
Website: www.caseheritage.org
Email: caseheritage@aol.com

ANTIQUE CATERPILLAR MACHINERY
Owners Club
7501 N. University, Suite 119
Peoria, Illinois 61614, USA
Telephone: 1-309-691-5002
Website: www.acmoc.org
Email: cat@acmoc.org

CHAMBERLAIN
Chamberlain 9G Tractor Club, Australia
Telephone: (08) 9523 0122
Website: www.chamberlain9g.org.au
Email: chamberlain9g@westnet.com.au

DAVID BROWN
David Brown Tractor Club
Box 990, Holmfirth, Huddersfield
West Yorkshire HD9 1YH, UK
Telephone: 01484 685875
Website: www.dbtc.co.uk

The David Brown Tractor Club of New Zealand
Website: www.davidbrown.org.nz

FERGUSON
Ferguson Enthusiasts of North America
5604 Southwest Road
Platteville, Wisconsin 53818, USA
Website: www.fergusontractors.org

The Friends of Ferguson Heritage
Nelson House, 2 Hamilton Terrace
Leamington Spa CV32 4LY, UK
Telephone: 01444 414087
Website: www.fofh.co.uk

FIELD MARSHALL
The Marshall Club
Briardale, 37 Scarborough Road, Norton, Malton
North Yorkshire YO17 8AA, UK
Website: www.themarshallclub.com
Email: bridget@themarshallclub.com

FORD
Ford N Tractor Club, USA
Website: www.ntractorclub.com

FORDSON
Fordson Collectors Association, USA
Website: www.ford-fordson.org

HART-PARR & OLIVER
Hart-Parr Oliver Collectors Association, USA
Telephone: 1-731-645-4769
Website: www.hartparroliver.org

INTERNATIONAL HARVESTER
IH Collectors Club Worldwide
310 Busse Highway, PMB 250
Park Ridge, Illinois 60068-3251, USA
Telephone: 1-847-823-8612
Website: www.nationalihcollectors.com
Email: ihcclub@aol.com

JOHN DEERE
Two-Cylinder Club
Box 430, Grundy Center, Iowa 50638-0430, USA
Telephone: 1-319-824-6060
Website: www.twocylinder.com
Email: two-cylinder@two-cylinder.com

MASSEY
Twin Power Heritage, CA
Website: www.masseycollectors.ca

Massey Collectors Association, USA
Website: www.masseycollectors.com

MINNEAPOLIS-MOLINE
Minneapolis-Moline Collectors
18581 600th Avenue, Nevada, Iowa 50201, USA
Website: www.minneapolismolinecollectors.org

NUFFIELD
Nuffield & Leyland Tractor Club, UK
Telephone: +44 (0)1453 828737
Website: www.thenuffieldandleylandtractorclub.co.uk
Email: contact@thenuffieldandleylandtractorclub.co.uk

RUMELY OIL PULL
Rumely Products Collectors, 5455 Elizabethtown
Road, Palmyra, Pennsylvania 17078, USA
Telephone: 1-717-917-8659
Website: www.rumelycollectors.com

REFERENCES

Magazines
Antique Power (USA)
Engineers & Engines (USA)
Old Tractor (UK)
Vintage Tractor & Country Heritage (UK)

Websites
www.canadianantiquetractor.com (CA)
www.oldtractor.come2me.nl (NL)
www.smokstak.com (USA)
www.steel-wheels.net (UK)
www.tractordata.com (USA)
www.tractordata.uk (UK)

Publications
Erb & Brumbaugh, *Full Steam Ahead*, American Society of Agricultural Engineers, 1993.

Farnworth, John, *A World-Wide Guide to Massey-Harris, Ferguson and Early Massey-Ferguson Tractors*, Japonica Press, 2000.

Tyler & Haining, *Ploughing By Steam*, Model & Allied Publications, 1970.

Stephenson, James, *Tractor Farming & Tractor Engineering*, Drake & Co., 1917.

Wendel, C.H., *Standard Catalog of Farm Tractors, 1890 to 1990*, KP Books, 2005.

Williams, Michael, *Massey Ferguson Tractors*, Blandford Press, 1987.

Williams, Michael, *Tractors of the World*, Parragon Publishing, 2005.

ACKNOWLEDGEMENTS

Thank you to the following individuals for their kind assistance: Chris Clemens, Stuart Gibbard, Guy Heaslip, Harold and Pansy Kent, Harold Kuret, Jeorg Mueller, David Parfitt, Brenda Stant and The Elite Steam Restoration Team.

PUBLISHER'S ACKNOWLEDGEMENTS

We would like to thank Paul Rackham, Liz Briggs and Lee Martin for their help and cooperation in arranging the photo shoot.

INDEX

A
Advance Rumely, Oil Pull, Model
  20-35 M 68, **69**
Allis-Chalmers (A-C) 9, 14
    Model B 20, **21**
    Model U 26, **27**
Austin, Herbert 28

B
British Wallis 92, **93**

C
cab(s) 14, 54
Case, Model CC-3 90, **91**
Caterpillar, Model D2 (5J) 30, **31**
Chamberlain, Model Super 70 94, **95**
crawler tractors 30, 58, 62, 72, 84, 88

D
David Brown 18
    Model 1210 86, **87**
    Model VAK 1 18, 44, **45**
Deutz-Allis 26
Diesel, Rudolph 11
Dreyfuss, Henry 22

F
Ferguson, Harry 7, 18, 24, 48, 64
Ferguson, Model TE-20 18, 24, **25**, 48
Ferguson-Brown, Model A 18, **19**, 44
Field Marshall, Model Series 3A 58, **59**
First World War 13, 30, 36, 62
Ford, Henry 13, 18, 24, 38, 48, 60
Ford, Model 9N 24, 48, **49**
Fordson 13, 28, 60
    Major, Model E27N 34, **35**, 38
    Model E1A-DKN 66, **67**
    Model F 36, **37**, 38

Model N 36, 38, **39**
four-wheel drive (4WD) 14, 56, 66, 86
Fowler-Marshall Ltd 58
Fowler, Model Challenger III 72, **73**
French Austin, Model R 28, **29**
Froelich, John 11, 76

G
garden-type tractors 15
Great War *see* First World War

H
Hart Parr 11, 12, 42
    Model 18-36 G 74, **75**
Holt, Benjamin 30, 62
Holt, Model 75 62, **63**

I
internal combustion engine 11, 72
International Harvester Company
    (IHC) 11, 12, 14, 40
    Model 8-16 Junior 60, **61**
    Model T-20 Crawler 88, **89**
    Model W4 84, **85**
    Titan, Model 10-20 46, **47**

J
John Deere 9, 14, 76, 78
    Model 3130 32, **33**
    Model R 22, **23**

K
kerosene 11, 28, 46, 60, 68, 88, 90

L
Lanz Bulldog 58
    Model D7506 78, **79**
Loewy, Raymond 14, 84

M
McCormick, Cyrus 40
McCormick Deering, Model 10-20
    Gear Drive 40, **41**
Marshall, Sons & Co. 12, 58
Massey-Ferguson (M-F)
    Model 35X 64, **65**
    Model 135 96, **97**
Massey-Harris (M-H) 24
    General Purpose (4WD) 56, **57**
Minneapolis-Moline (M-M), Model
    GT 54, **55**
MOM tractors 36
Morris, Sir William 50

N
Nuffield, Model 3/42 70, **71**
Nuffield Universal, Model M3 50, **51**

O
Oldfield, Barney 26
Oliver 74
    Model 80 Standard 42, **43**
orchard tractors 82, 88
Otto, Nikolaus 11

P
petrol 11, 20, 28, 50, 84, 88
ploughing match 102–7
pneumatic tires 26, 84
prairie tractors 12, 15, 60, 74
propane 54

R
Rackham, Paul 7, 9, 100–1
restoration 15
Royal National Lifeboat Institution
    (RNLI) 72

S
Saunderson, Herbert 12, 80
Saunderson, Model G Universal 80, **81**
shows 15, 108–9
Second World War 14, 15, 38, 50, 52,
    72
Society of Industrial Design 14
steam engines 10, 72

T
Thomas W. Ward Ltd. 58, 72
threshing 10, 11, 13, 34, 42, 54, 68,
    74, 76
tractor pulls 15, 22
Turner, Yeoman of England Mark 2
    52, **53**

U
unit frame construction 13, 28, 42

W
Ward Ltd., Thomas W. 58, 72
Waterloo Boy, Model N 76, **77**
Watt, James 10
Weeks-Dungey, Model New Simplex
    82, **83**
wheatland tractors 54
Winnipeg Motor Competitions 12
World War I *see* First World War
World War II *see* Second World War